AF269679

ENGINEERED BY NATURE

GREAT BARRIER REEF

BY MARTHA LONDON

CONTENT CONSULTANT
MIKHAIL V. MATZ
PROFESSOR
DEPARTMENT OF INTEGRATIVE BIOLOGY
UNIVERSITY OF TEXAS AT AUSTIN

Kids Core
An Imprint of Abdo Publishing
abdobooks.com

abdobooks.com

Published by Abdo Publishing, a division of ABDO, PO Box 398166, Minneapolis, Minnesota 55439. Copyright © 2021 by Abdo Consulting Group, Inc. International copyrights reserved in all countries. No part of this book may be reproduced in any form without written permission from the publisher. Kids Core™ is a trademark and logo of Abdo Publishing.

Printed in the United States of America, North Mankato, Minnesota
042020
092020

Cover Photo: Rich Carey/Shutterstock Images
Interior Photos: Shutterstock Images, 4–5, 6, 27; Rich Carey/Shutterstock Images, 7, 24; Red Line Editorial, 8, 28–29; Greg Sullavan/iStockphoto, 10; GSFC/LaRC/JPL, MISR Team/NASA, 12–13; Timothy Baxter/Shutterstock Images, 15; Coral Brunner/iStockphoto, 16 (foreground), 16 (background); Mikhail V. Matz, 18; Debra James/Shutterstock Images, 20–21; Richard Whitcombe/Shutterstock Images, 23; JC Photo/Shutterstock Images, 29

Editor: Marie Pearson
Series Designer: Megan Ellis

Library of Congress Control Number: 2019954277

Publisher's Cataloging-in-Publication Data

Names: London, Martha, author.
Title: Great Barrier Reef / by Martha London
Description: Minneapolis, Minnesota : Abdo Publishing, 2021 | Series: Engineered by nature | Includes online resources and index.
Identifiers: ISBN 9781532192876 (lib. bdg.) | ISBN 9781098210779 (ebook)
Subjects: LCSH: Great Barrier Reef (Qld.)--Juvenile literature. | Natural monuments--Juvenile literature. | Coral reefs and islands--Australia--Juvenile literature. | National parks and reserves--Juvenile literature. | Landforms--Juvenile literature.
Classification: DDC 910.202--dc23

CONTENTS

Snorkeling is a great way
to see the Great Barrier
Reef's beauty.

CHAPTER 1

AN UNDERWATER WORLD

The swimmer breathes through her snorkel. She looks around at the colorful Great Barrier Reef below. The sun is warm on her back. Small waves push against her body.

Ships look tiny compared to the Great Barrier Reef.

Below her, the reef is alive. Fish in every color dart back and forth. A red sea star lies on a rock. Pink and orange corals fan out. Anemones sway in the current. A tiny clown fish sticks its head through an anemone's **tentacles**. The reef seems to go on forever.

The Great Barrier Reef is in the Pacific Ocean. It is off the coast of Queensland, Australia. The reef is more than 1,400 miles (2,250 km) long. It is the largest collection of coral reefs in the world.

Sea anemones may look like plants, but they are actually animals.

Parts of a Coral Polyp

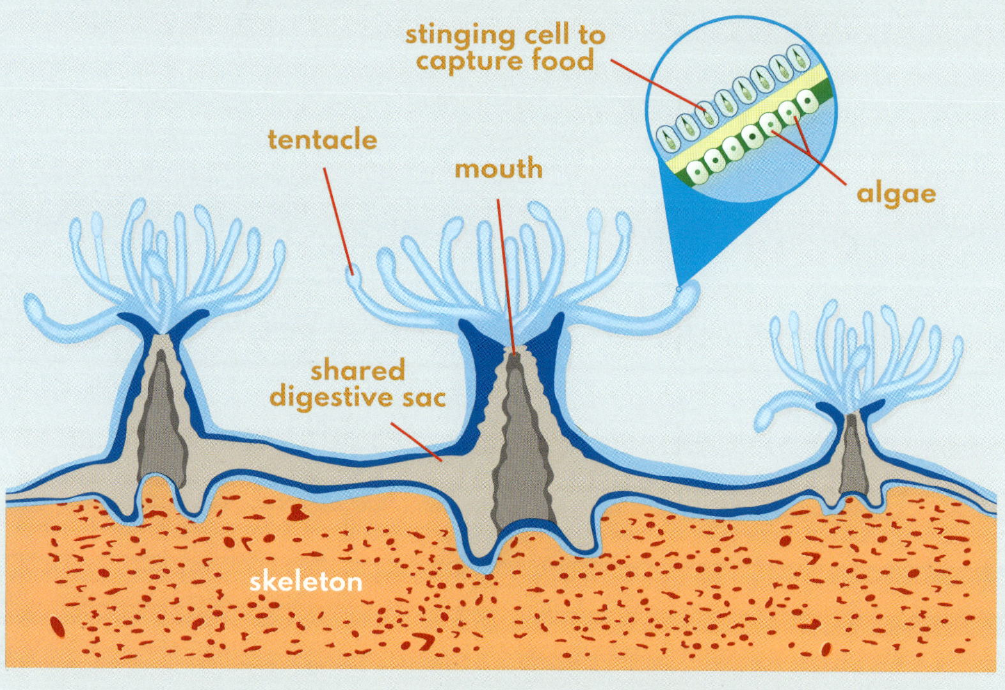

A shared digestive sac connects polyps, or individual corals, in a colony. The polyps make a skeleton under them. So unlike humans, a coral skeleton is actually outside the animal's body.

A Living Home

Millions of corals make up the reef. A coral is a type of animal. It is related to sea anemones. Corals and algae work together. Algae live in the corals. Algae get energy from sunlight.

They produce food for the corals. Algae give corals their dark color.

There are hard corals and soft corals. Both look a lot like plants. Hard corals have a limestone skeleton. Young corals grow on top of the older dead corals. They add more and more limestone to the reef. Over hundreds of years, the reef spreads and grows.

A Rainbow of Corals

The Great Barrier Reef has more than 400 **species** of corals. Corals are usually brown. This color comes from algae living inside them. Some corals have brighter colors, including green, blue, purple, and pink. These bright colors protect both corals and algae from the sun's harmful rays.

Sea turtles are some of the many animals that live
in the Great Barrier Reef.

Soft corals do not build reefs. But they do live on hard corals that died. The hard skeleton gives soft corals a good anchor.

The Great Barrier Reef is home to thousands of animal species and hundreds of plant species. This complex **ecosystem** is delicate. The Great Barrier Reef needs to be cared for so people and animals can continue to enjoy it.

Further Evidence

Look at the website below. Does it give any new evidence to support Chapter One?

How Great Is the Great Barrier Reef?

abdocorelibrary.com/great-barrier-reef

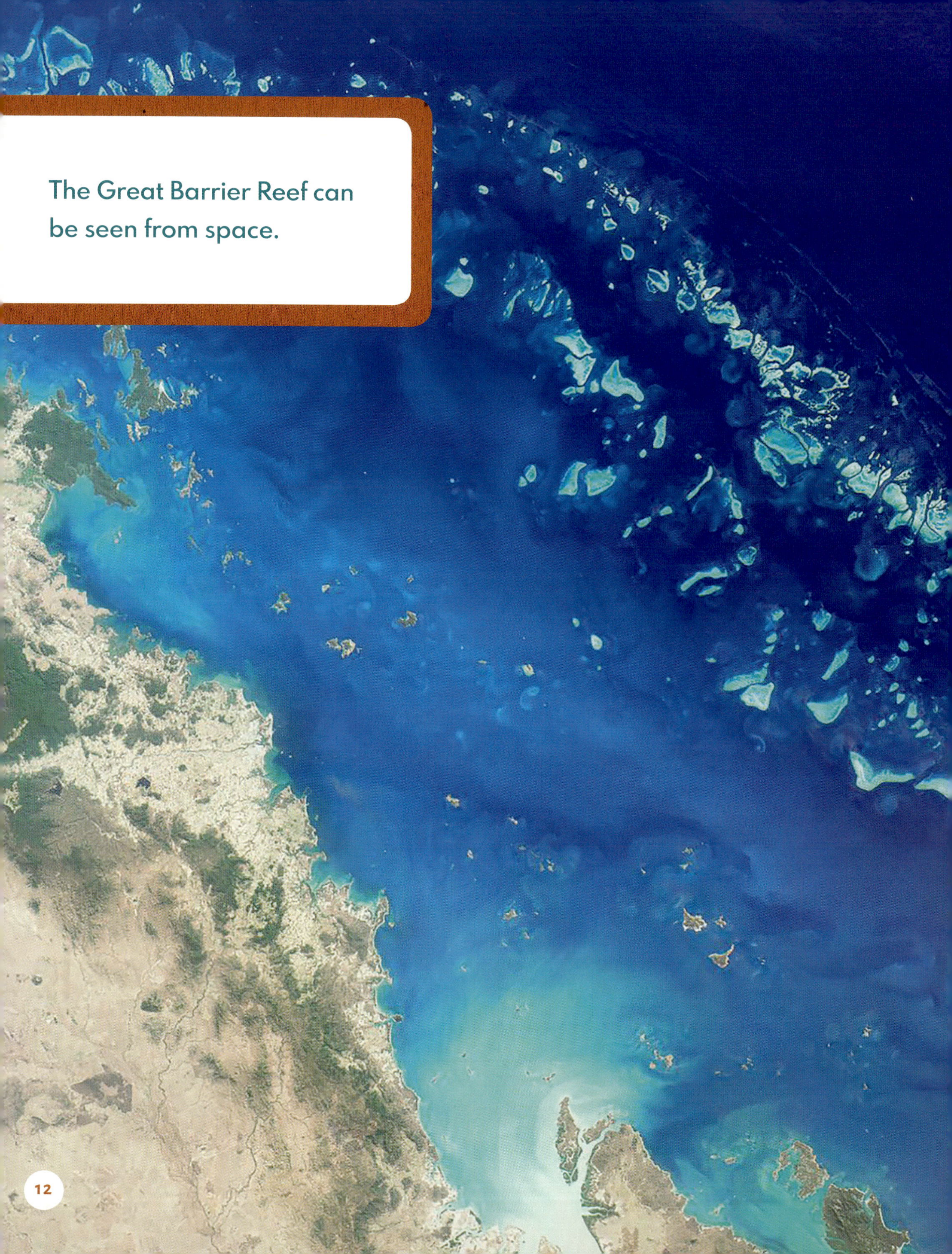

The Great Barrier Reef can be seen from space.

CHAPTER 2

A DELICATE BALANCE

The Great Barrier Reef began forming 500,000 years ago. But the reef that people recognize today is much younger. The ocean has changed during its existence. Sea levels have risen and fallen. Water temperatures have gotten hotter and colder.

Each time the ocean changed, the reef died. The reef later reformed. The current Great Barrier Reef is about 8,000 years old.

A Slow Process

The Great Barrier Reef continues to grow. However, corals grow slowly. Corals that create branches may grow up to 8 inches (20 cm) per year. Massive corals, such as brain corals, grow less than 1 inch (2.5 cm) per year. These corals form domes rather than branches.

Corals only spawn, or reproduce, once a year. During spawning, corals release special **cells** called eggs and sperm into the ocean.

The corals continue to change the reef.

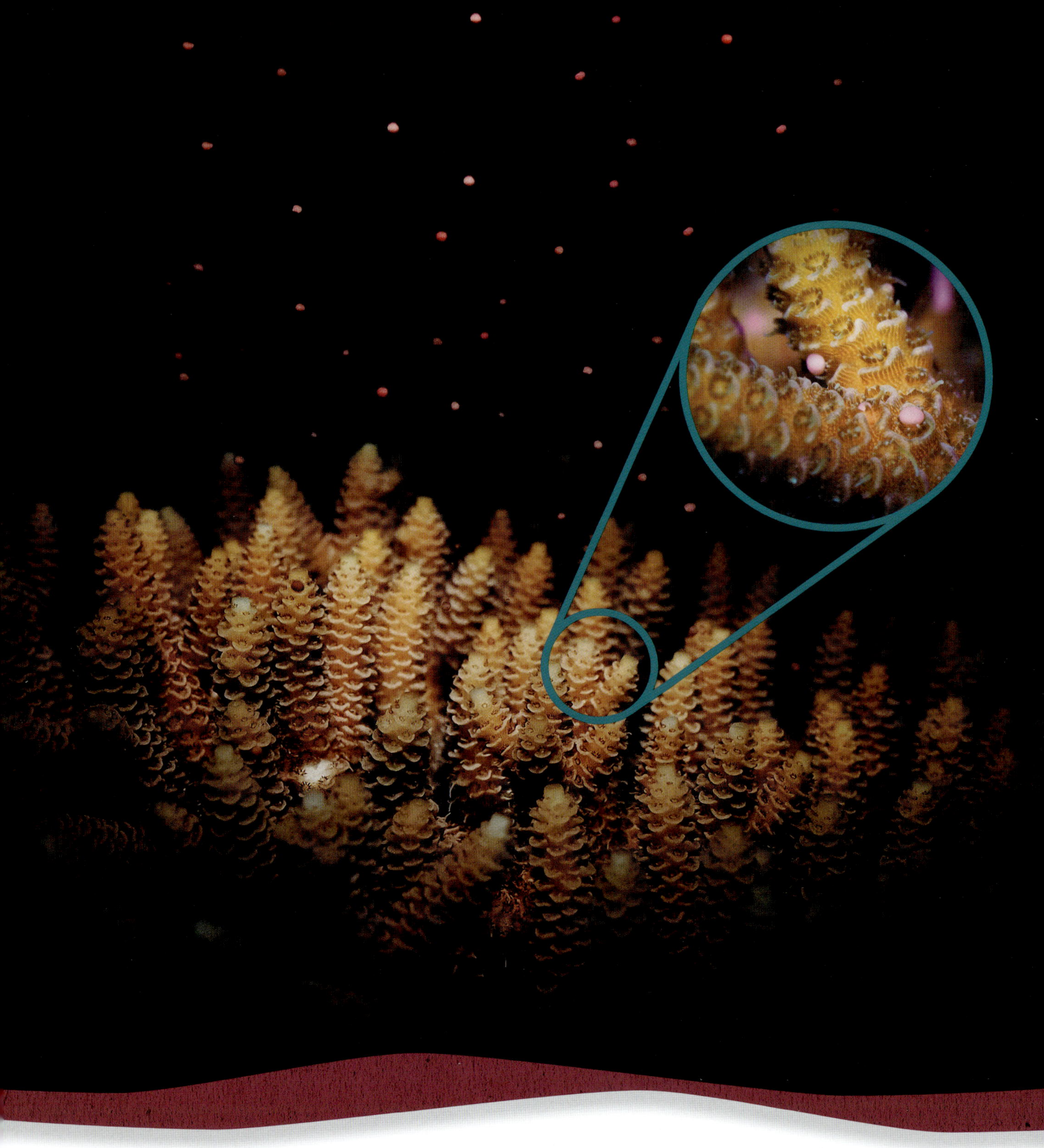

Corals release bundles of eggs and sperm, *pink*.

All corals of the same species spawn at the same time.

When an egg and sperm meet, they create a coral baby, called a larva. The larva is soft and tiny. It floats in the water. Millions of them travel with ocean currents for many days until they meet a hard surface to attach to.

Many Animals

More than 1,500 species of fish live in the Great Barrier Reef. They come in all shapes and sizes. Some butterfly fish grow to be 8 inches (20 cm) long. Groupers are large. They can be more than 8 feet (2.5 m) long. Groupers can weigh more than 800 pounds (360 kg).

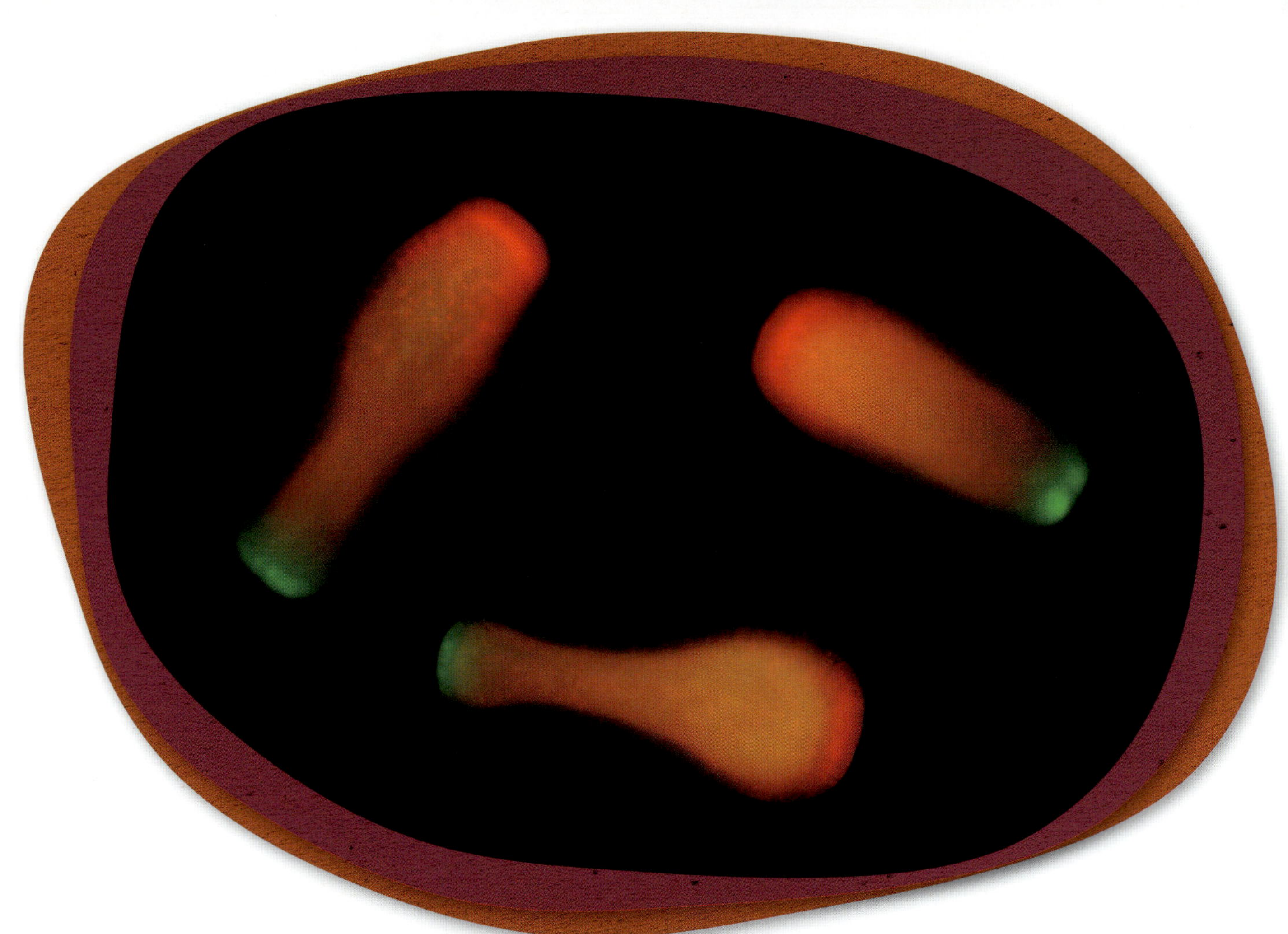

Coral larvae float in the ocean. The mouth is at the green end.

Spawning happens at night. It is usually one week after a full moon. At the Great Barrier Reef, spawning occurs in late spring or early summer.

Peter Mumby is a scientist. He explained in an interview that the reef's size helps keep it alive:

> The Great Barrier Reef is about the size of Italy and at any given time there are patches that have been damaged and patches that are pretty good, so it has an ability to heal itself.

Source: Helen Davidson. "Great Barrier Reef." *Guardian*, theguardian.com, 28 Nov. 2017. Accessed 1 Oct. 2019.

What's the Big Idea?

Read this quote carefully. What is its main idea? Explain how the main idea is supported by details.

Rising ocean temperatures put the animals and plants in the Great Barrier Reef at risk.

CHAPTER 3

IN THE FUTURE

Global warming is the heating up of Earth over time. Humans cause global warming. Burning fuel releases gases that trap heat in Earth's atmosphere. One of the effects of global warming is rising ocean temperatures.

The Great Barrier Reef is sensitive. Warming oceans harm corals. As temperatures continue to rise, the reef has begun to die.

Losing Its Color

Corals cannot live in water that is too warm. The algae leave the corals. This causes the corals to bleach. Bleaching is when the corals lose their normal brown color and become white.

When the algae leave, the corals lose their main source of food. The corals begin to starve. If the water does not cool, the algae will not return. The corals will die.

Bleaching makes corals more likely to get sick.

All fish are affected when corals die.

When corals die, animals have to find a new place to live. Corals build homes and provide food for thousands of animals. Every living thing in the reef is affected when corals die. Coral reef ecosystems need all animals to be balanced in order to remain healthy.

Healing the Reef

Many people are working to protect the Great Barrier Reef. Without the reef, the animals that depend on it would die. One of the ways people are working to protect the reef is by keeping the water clean.

Tourism

Every year, more than 2 million people visit the Great Barrier Reef. Money raised from tourism helps keep the reef healthy. The Great Barrier Reef Marine Park also keeps the reef safe. It does not allow boats near areas that could be easily damaged.

People in Australia are fixing shorelines.
Erosion of shores washes sand and dirt into
the reef. People grow grasses and other plants
along the shores. These plants keep sand and
dirt from washing away during rainstorms.
Farmers are also working to make sure fertilizer
does not end up in the ocean. Fertilizer can
hurt corals.

But the biggest work to be done is stopping
global warming. Scientists are working on

Explore Online

Visit the website below. Did you learn
any new information about coral reefs
that wasn't in Chapter Three?

Coral Reef

abdocorelibrary.com/great-barrier-reef

Plant roots help hold sand and soil together.

solutions such as using wind and solar energy.
These energies would replace burning fuel.
With hard work, the Great Barrier Reef and
the animals that depend on it will continue
to **flourish**.

MAP

- The Great Barrier Reef has more than 400 species of corals.

- The Great Barrier Reef is more than 1,400 miles (2,250 km) long.

- Approximately 1,500 species of fish make the Great Barrier Reef their home.

Great Barrier Reef
N
W
E
S
GREAT BARRIER REEF
PACIFIC OCEAN
29

Glossary

cells

the smallest parts of a living thing that are able to perform tasks needed for life

ecosystem

a community of plants and animals and their interactions with one another and their surroundings

erosion

the process of breaking down rock and moving sediment to a new area

flourish

to do well and grow in a strong, healthy way

species

types of plants and animals that are similar and able to breed together

tentacles

the flexible limbs of an animal

Online Resources

To learn more about the Great Barrier Reef, visit our free resource websites below.

Visit **abdocorelibrary.com** or scan this QR code for free Common Core resources for teachers and students, including vetted activities, multimedia, and booklinks, for deeper subject comprehension.

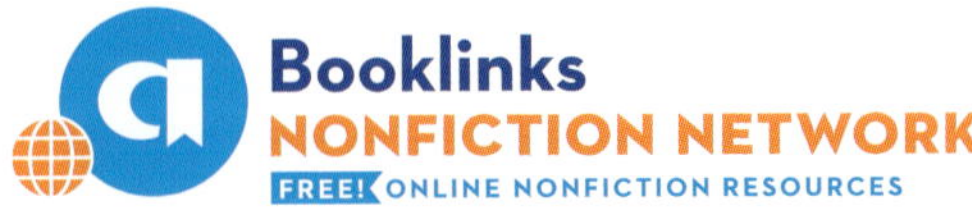

Visit **abdobooklinks.com** or scan this QR code for free additional online weblinks for further learning. These links are routinely monitored and updated to provide the most current information available.

Learn More

Hansen, Grace. *Marine Biome*. Abdo Publishing, 2017.

Medina, Nico. *Where Is the Great Barrier Reef?* Penguin Random House, 2016.

Index

About the Author

Martha London writes books for young readers full-time. When she isn't writing, you can find her hiking in the woods.